AF593573

A Stepping-Stone Book

THE LONG AND SHORT OF MEASUREMENT

By Vicki Cobb

Illustrated by Carol Nicklaus

Parents' Magazine Press—New York

Library of Congress Cataloging in Publication Data

Cobb, Vicki.
The long and short of measurement.

(A Stepping-stone book)
SUMMARY: Explains the principles of measurement and describes various measuring instruments.
1. Physical measurements–Juvenile literature.
[1. Physical measurement. 2. Measuring]
I. Nicklaus, Carol, illus. II. Title.
QC39.C573 530'.8 72-4903
ISBN 0-8193-0628-2 (lib. bdg.)

Printed in the United States of America

CONTENTS

ONE Sizing Things Up 5

TWO A Sense of More and Less 11

THREE From Here to There 16

FOUR Lengths for Everyone 23

FIVE Weight 34

SIX Temperature 41

SEVEN Time 47

EIGHT Measurement Is How to Know 57

Index 63

WORLD TRADE CENTER

Chapter One
SIZING THINGS UP

Some things in the world are very, very big. The tallest mountain is Mount Everest. It is as high as most jets fly. The World Trade Center in New York City is now the tallest building. For forty years, the Empire State Building, also in New York City, was the tallest. The largest ocean is the Pacific. The longest river is the Nile in Africa. The river with the most water is the Amazon in South America. Elephants and whales are the largest animals.

Can you think of some other big things? What is the biggest lake or river or building near where you live?

There are some things that are very small. The smallest furry animal is the musk shrew. It looks something like a rat and is about as long as your

thumb. The smallest bird is a hummingbird that comes from Cuba. It is even smaller than the musk shrew.

There are many things that are so small you cannot see them unless you use a magnifying glass or a microscope. If you want to see some small things, take a magnifying glass around your room. Look at a speck of dust, a crystal of salt, and a tiny piece of chalk powder.

What are some other small things you can find?

It is easy to tell the difference between the size of a mountain and a hummingbird. You know which is bigger than the other even when you cannot see them both at once. But sometimes it is hard to tell which one of two things is bigger even when they are side by side. If you have ever had to choose a piece of candy from a box full of candy, you know how hard it can be!

It is because things come in different sizes that we *measure* them. Measurement is used to answer

questions like: How long? How short? How high? How wide? How heavy? How hot? How fast?

Measurement is used to tell which is the biggest and which is the smallest. And measurement can be used to tell which is the bigger and which the smaller of two objects that are almost the same size.

Measurement is important for some people when they work. A construction worker must know measurement to build houses and bridges.

Grocers and butchers need measurement to know how much food they are selling.

Scientists use measurement to learn about nature. Their measurements told us which was the tallest mountain and the smallest animal.

People use measurement in their daily lives. When they shop, they tell storekeepers how much of something they want to buy.

When they travel, they want to know how far they are going.

When they know how hot or cold it is outside, they know how to dress for the day.

When they cook, they have to know how to measure different amounts of food in order to follow a recipe.

We can make appointments with a doctor, or a teacher, or a friend because we know how to measure time.

Different things are measured in different ways, but it is important to know many kinds of measurement.

Chapter Two
A SENSE OF MORE AND LESS

You get information about the world through your senses. If you want to know if a piece of furniture will fit through a doorway, you look at the width of the door and the width of the piece of furniture. You use your sense of sight to decide which is wider.

If you are asked to carry groceries, you might pick up one bag and then another. You use the senses in your muscles to decide which is heavier.

When you are going swimming, you stick your foot into the water. You use your sense of touch to decide how cold the water is.

Whenever you decide longer and shorter, or heavier and lighter, or warmer and cooler, you are *comparing* two objects. Comparing things is a rough kind of measurement. Your senses can tell differences between objects when the differences are large. But when the difference is small, your senses may not be able to detect it.

Compare the weights of a penny and a dime. Lift one, then the other. Can you tell which one is heavier? If you cannot, you can find out later on in this book how to measure the difference.

Your senses can also be fooled. Can you tell, by looking, which of these two long lines is longer?

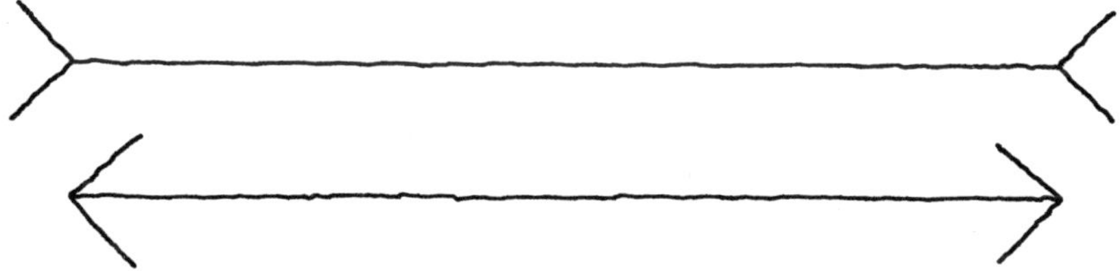

Get a small paper bag and a large paper bag. Put a jar of jelly or some small can in the small paper bag and fold the top so you cannot see inside. Fill the large paper bag with crumpled newspaper and fold down its top. Ask a friend to lift first one and then the other and tell you which one is heavier. Chances are your friend will think that the larger bag is heavier even when it is not.

Your sense of touch can be easily fooled. Get three bowls large enough to put your hand in. Put very warm water in one bowl, but be sure the water is not too hot to touch. Put cold water in another bowl, and put lukewarm water in the third bowl. Put one hand in the cold water and one in the very

warm water. Hold them there while you count to 100 slowly. Then put both hands in the lukewarm water. How does the lukewarm water feel to each hand?

People found out long ago that their senses could not tell small differences. They discovered that their senses can be fooled. They found that even when they could decide longer and shorter, or heavier and lighter, or hotter and colder, they could not tell exactly how long or how heavy or how hot something was. And they learned that the sense of sight was better at telling differences than any of their other senses.

So people invented measuring tools. There are measuring tools to tell how long things are, how heavy they are, and how hot they are. When you use a measuring tool, you usually use your sense of sight. Measuring tools make it easier to see small differences and make sure you are not fooled. In order to learn how to measure, it's a good idea to understand how a measuring tool works.

Chapter Three

FROM HERE TO THERE

One thing we measure is *distance* or *length.* Distance is a line you can draw, or walk, or travel, or imagine from one spot to another. Distance can be a straight line or a line with curves and corners. It can be as long as from the North Pole to the South Pole. It can be as short as the length of an eyelash or the letter "l." It can even be as large as the universe or so small that it cannot be seen through the most powerful microscope.

If someone wanted to give you an idea of how long a certain distance was, he might say, "It's as long as my arm" or "It's about a block long." In your imagination, you compare the distance you don't know with a distance you do know.

But you can get an even better idea of how long something is by seeing lengths side by side. You can, for example, compare the lengths of other objects with the length of a pencil by placing them alongside the pencil. You have to be careful that one end of the object is even with one end of the pencil. If the other ends are even, you know that the length of the object is the same as the length of the pencil.

Compare the length of a pencil with objects around your house or school. Is it the same length as a toothbrush, or a fork, or a chalkboard eraser?

You will probably have a hard time finding objects that are just as long as the pencil. Almost everything will be longer or shorter than the pencil. Perhaps your toothbrush is a little bit shorter than the pencil. But how much is a little bit? Here is a way you can get a better idea of just how long something is.

Suppose you want to measure the length of your pencil. Get some paper clips that are all the same size. Place the end of one paper clip so that it is even with one end of the pencil. Then make a straight line of paper clips, end to end, alongside the pencil.

You may find that your pencil is just as long as a certain number of paper clips. But the chances are that your pencil is longer than, say, 6 paper clips but shorter than 7 paper clips. In other words, you can say your pencil is between 6 and 7 paper clips long.

The length of your pencil is a count of lengths that are all the same size. A paper clip is your *unit of measurement.*

You can use many different things as units of measurement, even parts of your body. You can measure the length of a room with your feet. Put the heel of one foot against a wall of your room. Walk a straight line to the wall you are facing by putting the heel of one foot against the toes of the other. Count how many steps you take this way to reach the opposite wall.

Suppose you needed 30 steps to reach this point and that there is a small space left over between your foot and the wall. This space is smaller than your foot. You can now say, "The room is longer than 30 of my foot lengths but shorter than 31 of my foot lengths."

You can measure other distances in your foot lengths the same way. When you compare these measurements you can say *how much* longer or shorter they are than the length of the room.

Compare the length of your room to the length of the walk leading up to your school, or to another room in your house, or to a block in your

neighborhood. Make some measurements with other units of measurement such as uncooked spaghetti or pennies. Be sure to lay your units in a straight line with the ends touching each other.

You can measure distance in different directions. The word length means distance, but it can also mean the longest side of something. The longest side of a room, or a box, or a rug is its length. Sometimes it is hard to decide which side is longer.

A direction that makes a corner with the length is the *width* of an object. When you turn a corner and measure the shorter side of a room, or a box, or a rug, you are measuring its width.

You can also measure distance in an up and down direction. This distance is called *height.* The height of a room is from floor to ceiling. The height of a box tells you how deep the box is. Height is from the bottom to the top or from the top to the bottom of an object.

When you know the length and width and height of a space or an object, you know its *dimensions.* Dimensions give you a very good idea of just how big something is.

Chapter Four

LENGTHS FOR EVERYONE

When people first started measuring length, they used units of measurement that were handy. What could be more handy than parts of their bodies? They stretched out an arm and called the distance from the tip of a finger to the tip of the nose a *yard.*

The distance from heel to toe was called a *foot.*

The distance across the palm of the hand was called a *hand.*

And for measuring small distances, they used the width of the thumb, which they called an *inch.* But there was a serious problem with these units of measurement. People come in different sizes. If everyone used parts of his or her own body to measure length, no one could be sure just how long something is.

Suppose a person wanted to buy some cloth by the yard. He would get more cloth from a tall salesman with long arms than from a short salesman.

Suppose you were building a clubhouse and you knew you needed a board 15 of your foot lengths

long. You tell the man at the lumber yard, "I need a board 15 feet long." Then the man goes off and measures a board using *his* feet as the measuring unit. You probably would not get the length you wanted.

So it was decided long ago that units of length had to be the same size for everyone. There is a story that the length of a foot is the same as the foot length of a king who lived long ago. How people decided what length to call a foot is not important. What *is* important is that everyone agreed to call a certain length one foot. It is a *standard* measuring unit. Whenever people say "one foot" everyone knows just how long that is.

A tool that can be used to measure length in feet is simply a stick one foot long. People often call such a stick a *ruler.* Every ruler one foot long is exactly the same length as every other foot ruler. You can get an idea of the size of a foot by comparing the length of such a ruler with objects you find around you. How many things can you find that are *exactly* one foot long? How many things can you find that are *about* one foot long?

If you collect a few rulers, you can use them to measure the top of a table. Put the first ruler on the table with one end against the edge of the table. Place a second ruler with one end against the end of the first ruler. Make sure there is no space between the two rulers. You want to measure every bit of space.

Keep on placing rulers, end to end, in a straight line, just as you placed paper clips when you measured your pencil, until you reach the other end of the table.

If the last ruler fits exactly, the length of the table in feet is a count of the number of rulers you used. If there is a space left over that is too small for another ruler, the table is longer than the number

of rulers you used. But it is shorter than one more foot length. It could be between 3 and 4 feet long.

If you cannot get enough rulers, you can cut strips of construction paper that are each one foot long and use them the same way. Check the length of each strip against a ruler to make sure it is exactly one foot long before you place it on the table.

A foot is not useful for measuring lengths smaller than one foot. So we use another standard unit of measurement called an *inch.* Instead of using many sticks each one inch long, we use a ruler to measure inches. A ruler one foot long is just as long as 12 inch sticks laid end to end. In other words, there are 12 inches in a foot. On a ruler, the inches are numbered, so you always know how far you are from the end of the ruler.

You can use a ruler one foot long to measure the length of the line on this page in inches. Place the end with the 1-inch mark so that it is even with the left end of the line. Read the number of the inch at the other end.

Measure other small things in inches. How long is a candy bar? How long is a pencil? How long is a piece of spaghetti?

How do you think the small spaces marked off between inch marks can help you measure lengths that fall between inch marks?

When people want to measure distances longer than one foot they may use another stick called a *yardstick.* A yardstick is the same length as 3 foot rulers laid end to end, or 36 inches.

You can use a yardstick to measure the length of your school yard. Since you may not be able to get enough yardsticks to lay end to end, you can measure the yard with only one measuring stick. Here is how to do it:

Place one end of the yardstick against the fence at one end of the playground. At the other end of the stick, mark the ground with a piece of chalk or a small stone. Pick up the yardstick and place the end that was against the wall at the mark. Make another mark at the other end. Keep placing the yardstick

P.S.42

so that you make a straight path to the opposite end of the yard.

Does the yardstick fit exactly? Can you measure any left-over distance in feet or inches? A count of the number of marks you made is a measure of the playground in yards.

About how many yards long is your playground? A football field is 100 yards long. Could a football game be played in your school playground?

Distances much longer than a football field are measured in *miles.* A mile is as long as 5,280 one-foot rulers laid end to end.

When you travel in a car or bus, miles are measured by a special instrument called an *odometer.* Odometers keep track of the miles by the number of times the wheels go around.

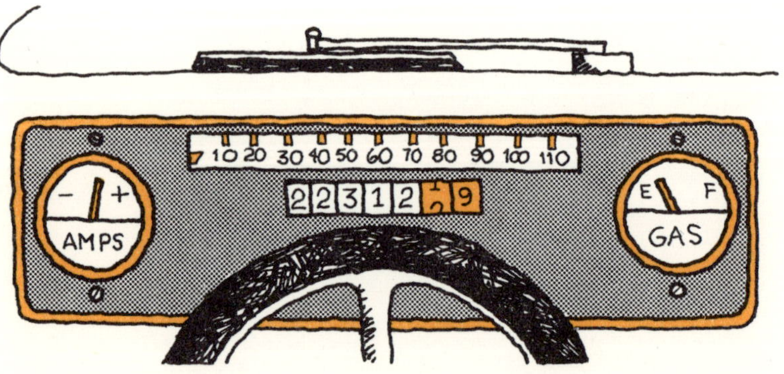

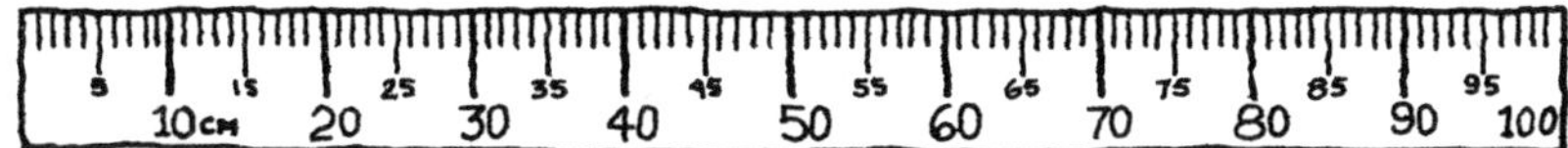

Americans measure distance in feet, inches, yards, and miles. These are standard measuring units in what is called the *English system.*

Scientists and most nations of the world use a different set of standard units for length that are part of the *metric system.* In the metric system, a *meter* stick is used instead of a yardstick. A meter is about three inches longer than a yard. A meter is marked off in 100 equal spaces. Each space is called a *centimeter. (Centi* means one hundred.) You use a meter stick the same way you use a ruler or yardstick.

Chapter Five

WEIGHT

Another thing people measure is weight. Weight is the pull of the earth on objects near its surface. You can feel weight when you lift objects or when an object is resting on you. But people don't use the senses in their muscles to measure weight.

If you sit on one end of a seesaw and no one sits on the other end, your end will rest on the ground and the other end will be in the air.

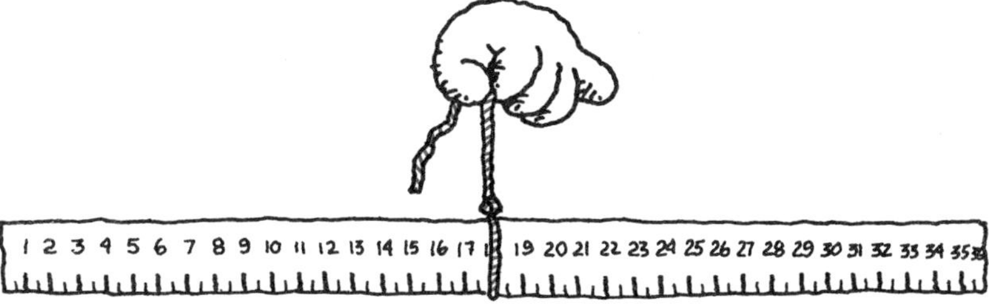

If you hang an object on a spring or a rubber band, the spring or rubber band stretches.

The changes weight makes in the balance of a seesaw or the length of a rubber band or a spring led to the invention of measuring tools for weight.

Weight-measuring tools are called *balances* and *scales.* Balances and scales tell you how heavy or light something is by measuring distance.

You can make a simple balance out of a measuring stick. Tie a string around the middle of a yardstick, at the 18-inch mark. Leave enough string so you can hold it, or tie the stick to some object. Make sure that the ends of the stick are free to swing up and down like a seesaw.

When the stick dangles freely, both ends should come to rest the same distance from the floor. In other words, the stick is balanced. If you hang an object on each end, the heavier object will be closer to the floor. When two objects have the same weight, the ends of the stick will be balanced.

Make handles on two paper cups, like the ones in the picture. You can now use your balance to weigh small things. Hang each cup one inch from each end. Your measuring stick should be balanced. Put a penny in one cup and a dime in the other. Which is heavier? Can you find some other small things to weigh?

There are many kinds of balances and scales. They all use units of measurement to tell just how heavy an object is.

Long ago when people weighed things, they used handy objects like cereal grains or pennies to measure weight. You can see how by doing as they did with your measuring-stick balance.

Put a small object like a spool of thread or a pair of scissors in one cup of your measuring-stick balance. Which side of the stick is now closer to the ground? Now add, one by one, small objects that are alike, such as pennies or paper clips, to the other cup until the stick is balanced. A count of the number of pennies or paper clips needed to balance the stick is the weight of the object.

Pennies and paper clips can be units of measurement for weight.

When people used handy objects for measuring weight, they had the same kind of problem that they had using their bodies to measure length.

Pennies, for example, come in different weights. An old, worn penny weighs less than a new one. So people finally decided to have standard units for weight.

A standard unit for weight is an object that is used only for weighing things. Weight units are usually made of metal that does not rust or change in any other way. A certain amount of sterling silver, for example, was called one *pound.*

Although pound weights might be made of any kind of metal, every pound weight would balance a pound of sterling silver.

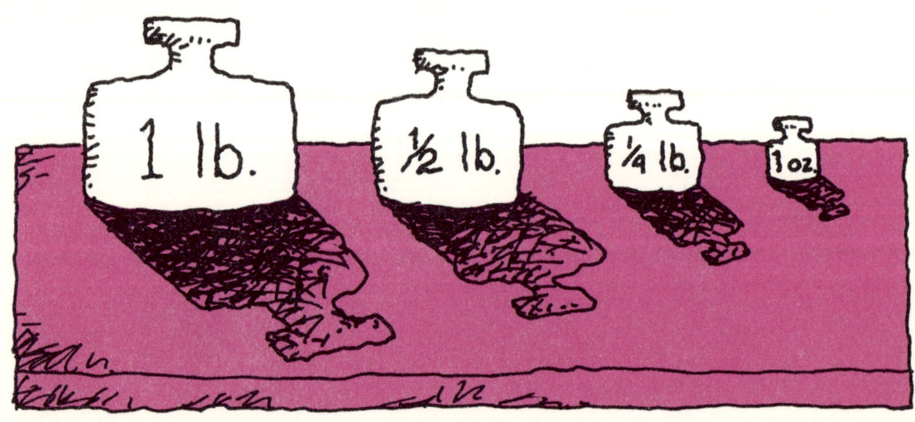

If you wanted to measure out a pound of hamburger, you would put the pound weight on one side of a balance and add hamburger to the other side until both sides were balanced.

Another standard unit of weight is an *ounce.* A small chocolate bar weighs about one ounce. There are 16 ounces in a pound, which means a 16-ounce weight balances a one-pound weight. Pounds and ounces are standard units in the English system.

As you might expect, the metric system uses a

different set of weights. A small weight is the *gram.* A gram is about the weight of a medium-sized paper clip. A *kilogram* is a larger weight unit in the metric system. There are 1,000 grams in a kilogram. One kilogram weighs a little more than 2 pounds.

Some people who measure weight in their work are butchers, druggists, and post-office clerks. You might like to find out what units of weight each one uses most—ounces, pounds, or grams.

Chapter Six
TEMPERATURE

A way of measuring hot and cold was invented long after people had figured out how to measure length and weight. Scientists first had to discover a way to *see* a difference between an object when it is hot and when it is cold. After much searching, they found that certain materials get bigger when they are heated and smaller when they are cooled. One of these materials is a silvery, liquid metal called *mercury*.

The changes heat makes in the size of mercury are used in a measuring tool that measures the heat of other materials. *Temperature* is a measurement of heat. A *thermometer* is the measuring tool.

A mercury thermometer is simply a hollow glass

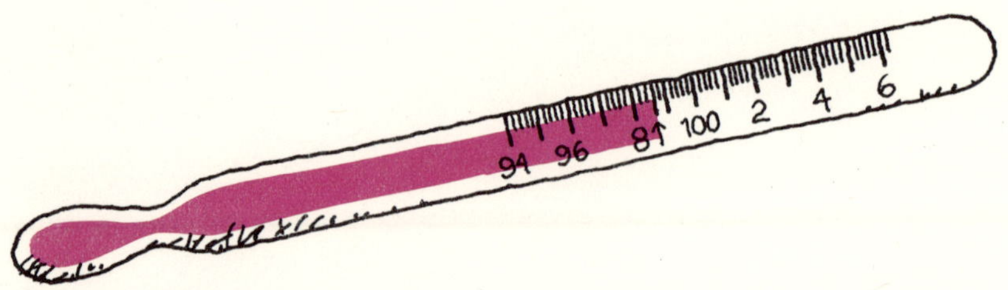

tube with a small bulb on one end. The bulb is filled with mercury. When the bulb of the thermometer is heated, the mercury grows larger. There is only one place for mercury to go as its size gets larger–up the tube. You see it through the glass as a silvery line. When the bulb is cooled, the line of mercury gets shorter. In other words, a thermometer measures temperature by length–the length of a mercury line.

All that is needed is some kind of ruler on the outside of the thermometer that can tell you the length of the mercury line. The markings on the outside of a thermometer are just like a ruler. It is marked off in tiny spaces that are all the same distance apart. Each space is called a *degree.* Every degree has a number just as inches and centimeters are numbered on a measuring stick.

When you use a thermometer, it is important to have the bulb completely surrounded by the material you are taking the temperature of. You can take the temperature of water, air, or your own body.

The length of the mercury line changes until the temperature of the mercury is the same as the temperature of the thing you are measuring. That is, the line stops moving. When this happens, you find the temperature by reading the number of the degree at the end of the mercury line.

There are standard measuring units for temperature, just as there are for length and weight.

First, scientists had to find something they could always count on to be the same temperature. They found that at sea level the temperatures of freezing water and boiling water never change. Every time you put a thermometer into boiling water, the length of the mercury line is the same. The same thing is true when you put a thermometer into ice water. Of course, the mercury line is much longer in the boiling water than it is in ice water.

The English system and the metric system have different thermometers.

The English system uses the Fahrenheit thermometer. On it, the point where the mercury line ends when water freezes is called 32 degrees. The boiling point is called 212 degrees. The distance between these points is divided into 180 equal spaces.

On the Fahrenheit thermometer zero degrees is

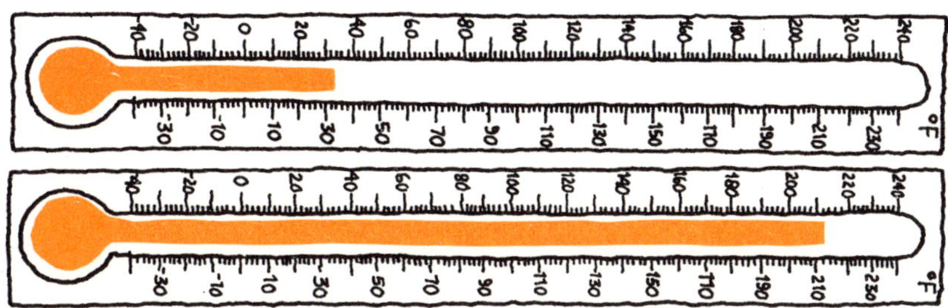

very cold. If it is zero degrees outside you have to dress very warmly. The temperature of your body is 98.6 degrees Fahrenheit, unless you are sick.

The metric system uses the Celsius thermometer. It is sometimes called the centigrade thermometer. The point at which water freezes is called zero degrees Celsius. The point at which water boils is called 100 degrees Celsius. On the Celsius thermometer; the distance between these two points is divided into 100 equal spaces.

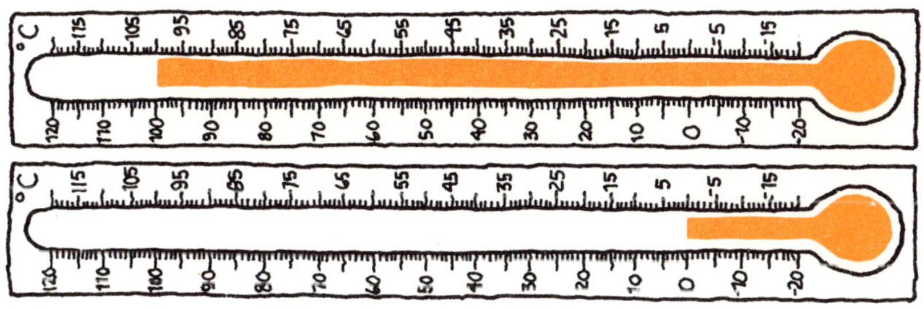

A day that is zero degrees Celsius is not as cold as a day that is zero degrees Fahrenheit. The temperature of the human body is about 36 degrees Celsius.

Measuring temperature is important for weathermen, bakers, and doctors. What do they measure the temperature of? Why must they measure temperature?

Chapter Seven
TIME

In order to measure time, people needed to find something that happened over and over again in a regular way. Long ago, people found what they needed to measure time by watching the sky. They saw the sun rise and set. In between sunrise and sunset you could draw an imaginary line that traced the path of the sun across the sky. Every day this line was in a slightly different place from where it was the day before.

At a certain time, at the beginning of summer, the path of the sun was as close to being directly overhead as it ever got. People noticed that it was a long time before the sun made this same path across the sky. They called this time a *year.*

At night people noticed that the shape of the

moon changed in a regular way. It started as a thin curved line—a *crescent*—and grew larger until it became a full circle. Then it grew smaller each night until you could not see it at all. Then it grew larger again until it was a full circle. This happened over and over again. A moon shaped like a circle is called a *full moon.* People called the time from one full moon to the next a *month.*

The time from sunrise to sunrise or from sunset to sunset was called a *day* by many people. But measuring a day by the rising or setting sun has its problems. In the fall the sunsets get earlier and earlier as days get shorter and shorter. In the spring the opposite thing happens.

Over two thousand years ago, the Egyptians figured out a better way to measure a day. You can do as they did. On a bright, sunny day, put one end of a stick in the ground so it points straight up. Make sure that the ground is level and that the stick casts a clear shadow. Lay a ruler or yardstick next to the shadow so you can measure its length. (You have made your own sun dial.)

You will find that during the morning the shadow gets shorter and during the afternoon the shadow gets longer. There is a moment, around noon, when the shadow is as short as it will get.

On winter days the shortest shadow is longer than the shortest shadow on summer days. But the

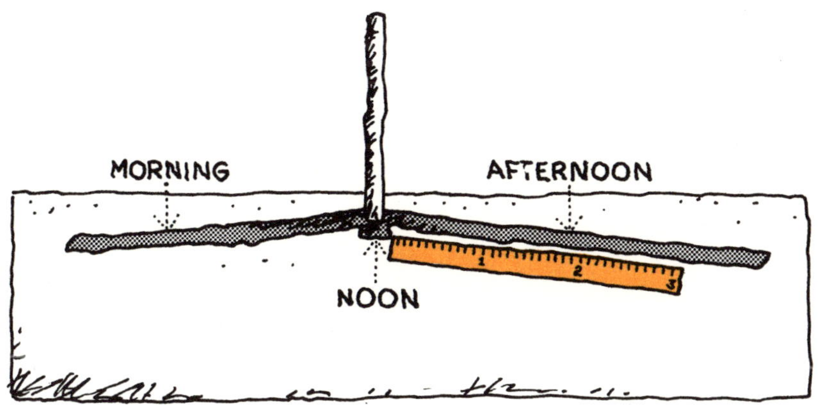

time between the shortest shadow on one day to the shortest shadow the next day is always the same no matter what the time of year.

The tool people invented to keep track of days, months, and years is the *calendar.* Calendars divide the year into 12 months. Every month has a name. Do you know the names of all the months of the year?

Every day of each month has a number. Four months have 30 days–April, June, September, and November.

One month, February, has 28 days except every fourth year when it has 29 days.

All the other months have 31 days. Can you name the months with 31 days?

Years are also numbered. What is the number of this year? The number of the day, the name of the month, and the number of the year is called a *date.* What is today's date?

Calendars make it possible to keep track of dates even when people cannot see the sky. Calendars make it possible to always know when important things happened, such as the day you were born. You can also know when something important is going to happen in the future. Do you know the date of something important coming up?

Clocks are another measuring tool for time that divides the day into shorter periods. Today's clocks and watches look very different from the first clocks people invented. One of the very first clocks was like the stick you put in the ground. The time of day was measured from the length of the stick's shadow. But all clocks, from the earliest ones to the ones we use today, divide the day the same way.

The face of a modern clock is usually a circle. The distance around this circle is divided into 12 even spaces and numbered from 1 to 12. There are at least two pointers attached to the center of the clock face. These pointers are called the *hands* of the clock.

The shorter hand moves so slowly you cannot see it moving. In one day, this hand makes two complete circles around the face of the clock. The time it takes this hand to move from one number to the next is called an *hour.* It points directly to a number 24 times during the day. In other words, there are 24 hours in a day. Each half day has 12 hours, and each hour in half a day has a number.

The hour is called o'clock, which really means "of the clock." And this smaller hand, called the *hour hand,* tells you whether it is 3 o'clock or 7 o'clock, or whatever number it is pointing to.

Many clocks show the distance between numbers divided into five equal spaces. There are 60 spaces around the face of the clock. The spaces each stand for a time called a *minute.*

The longer hand of the clock, called the *minute hand,* moves around the face of the clock much faster than the hour hand. In fact, it takes the minute hand only 1 hour to make a complete circle and point to every one of the 60-minute spaces.

It takes the minute hand 5 minutes to move from one number to the next.

The minute hand makes a complete circle around the face of the clock 24 times every day.

When you tell time, you use the minute hand to tell you how many minutes it is *past* the hour or how many minutes *to* the next hour. When the minute hand points to 12, then the time is the number that the hour hand is pointing to. It is 4 o'clock when the minute hand is on the 12 and the hour hand is pointing to 4. The picture shows the number of minutes past the hour and the number of minutes to the next hour for each number of the clock.

Many clocks have a long, thin hand that moves very quickly. This hand makes a complete circle around the face of the clock in 1 minute. It makes 60 trips every hour. Since it points to the 60 markings around the face of the clock every minute, these markings divide a minute into 60 equal times called *seconds.*

As you might expect, this hand is called the *second hand.* How many seconds does this hand take to move from the 2 to the 3 or from the 7 to the 8?

The hands of a clock measure time by moving at a steady rate through a distance–the distance around the face of the clock. Which hand moves the shortest distance during one day–the second

hand, the minute hand, or the hour hand? (Hint: which hand makes the fewest number of round trips?) Which hand moves the longest distance during a day?

The clock and the calendar are probably used by more people in their daily lives than any other measuring tools. Find out how many times during the day you want to know what time it is. It is important for television stations, airports, and hospitals to keep track of time. Can you think why? Where else is the measurement of time important?

Chapter Eight

MEASUREMENT IS HOW TO KNOW

Suppose someone said to you, "My dog weighs 30 pounds" or "That is a 20-inch bicycle wheel." You probably would not ask "How do you know?" You know how people find out pounds and inches. If you think there may be something wrong about a measurement, you can get the right measuring tool and check for yourself.

Scientists use many measuring tools. Most of them are more complicated than yardsticks and balances. Many tools measure things that we could not measure with our senses.

A scientist who studies weather has a tool to measure the speed of the wind and another to measure the weight of the air.

Scientists who study the stars have a tool that

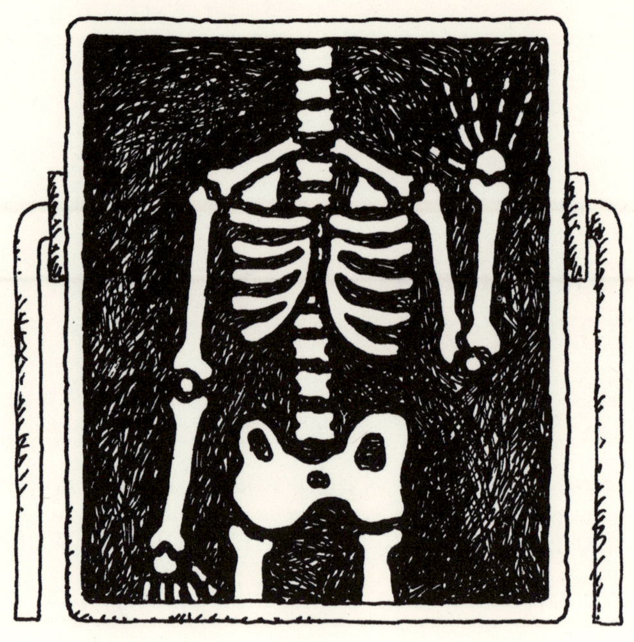

measures a kind of light, like x-rays, that is invisible to the human eye.

Scientists who study the materials of the earth have tools that measure the size of atoms—tiny bits of matter that are impossible to see even with a microscope.

Without these measuring tools, we would not know very much about wind speed, or the weight of air. We would not know that x-rays and atoms are real things.

Whenever scientists make a new discovery,

someone almost always asks them, "How do you know?" A scientist cannot answer, "I just believe it." He or she must be able to answer, "I will tell you what I did to find out. If you do what I did, then you will know what I know."

Through measurement, scientists share the excitement of their discoveries, and often other scientists find out more information about new discoveries.

You don't have to be a scientist to start inventing ways of measuring things. You can even use everyday measuring tools and standard units of measurement. You could measure how much rain fell during a storm by building a simple rain gauge. All you need is a can, such as a soup can, and a ruler. Stand the ruler straight up in the can.

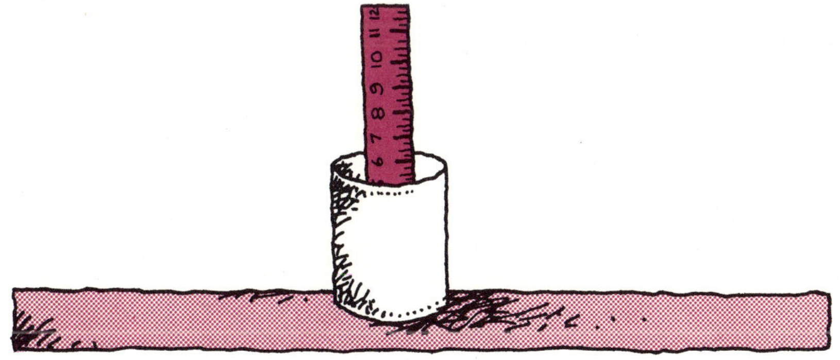

Leave the can and ruler outside during a storm. After the storm is over, measure how deep the water is. If you do this for many storms, you will find out that more water falls in some storms than in others.

You can find out if beans give off heat as they sprout if you use a dairy thermometer. Perhaps there is one in your school laboratory, or you can get one in a hardware store. You could then make this a classroom experiment.

Soak some dried lima beans overnight. The next day, drain the beans and put them in a container that does not let heat escape. A thick, lightweight, plastic coffee cup is good if you cover it with several thicknesses of cardboard.

Make a hole in the cover so you can stick in the thermometer. Make sure the bulb of the thermometer is surrounded by beans.

Every morning read the thermometer and then check the beans to see how they are sprouting. After a day or so, if the beans seem dry, rinse them by putting them in a strainer and letting water run over them. Return the rinsed beans to the container. Keep a record of your temperature reading. You will find that the temperature goes up as the beans sprout. They should sprout in about a week.

When you begin inventing your own ways to use measurement to learn about nature, you are thinking like a scientist.

INDEX

airports, and time, 56
Amazon River, 5
atoms, 58

bakers, and temperature, 46
balances, 35-39
beans, experiment with, 61-62
big and small, 5-8
board, measuring of, 25
bodies, measuring with, 19-20, 23-25
butchers, and measurement, 9;
and weighing, 40

calendar, 50-51, 56
Celsius thermometer, 45-46
centimeter, 33, 42
clocks, 51-56
cloth, measuring of, 24
cold and hot, 12, 13, 14-15, 41, 44
comparing, 13, 17, 20
construction paper, as ruler, 28
cooking, and measurement, 10

date, 51
day, 48, 50, 51;
division of, 51-56
degrees, 42, 43, 44-46
dimensions, 22
discoveries, scientists and, 58-59
distance, and weight, 35;
see also length
doctors, and temperature, 46
druggists, and weighing, 40

Egyptians, and time, 49
elephants, 5
Empire State Building, 5
English system, 33;
for weight, 39;
thermometer for, 44

Fahrenheit thermometer, 44-45, 46
feet, 33
foot, 23
foot lengths, 19-20, 24-25
football field, length of, 31

gram, 40
grocers, and measurement, 9

hand, 23
hands, of clock, 52-56
heavier and lighter, 12, 13, 14, 15, 35-36;
see also weight
height, 22
hospitals, and time, 56
hot and cold, 12, 13, 14-15, 41, 44
hour, 52-55
hour hand, 53
human body, temperature of, 45, 46
hummingbird, 6, 7

inch, 23, 29-30, 33, 42

kilogram, 40

length, 16-21, 22, 23-33;
and temperature, 42
light, measurement of, 58
lighter and heavier, 12, 13, 14, 15, 35-36
longer and shorter, 13, 15, 20

magnifying glass, 6
measurement, purposes of, 7-10;
tools for, 15, 35, 41, 50, 51, 57-58, 59;
unit of, 19, 21;
ruler for, 26-30;
of temperature, 46, 61-62;
inventing ways of, 59, 62
mercury, 41, 42, 43, 44
meter, 33
metric system, 33;
for weight, 39;
thermometer for, 44, 45

microscope, 6
miles, 32-33
minute, 53
minute hand, 53-54
month, 48, 50, 51
moon, shape of, 48
Mount Everest, 5
mountain, 5, 7
musk shrew, 5-6

Nile River, 5

o'clock, 53
odometer, 32
ounce, 39, 40

Pacific Ocean, 5
paper clips, as units of measurement, 18-19, 37;
experiment with, 36, 37
pencil, measuring with, 17-18;
measuring of, 19
pennies, as units of measurement, 21, 37-38
people, and measurement, 9-10
playground, measuring of, 30-32
post-office clerks, and weighing, 40
pound, 38-39, 40

rain, measurement of, 59-60
ruler, 26-30, 33;
and thermometer, 42

scales, 35, 37
scientists and measurement, 9, 57-59, 62;
discoveries of, 58-59
sea level, and temperature, 44
second hand, 55
seconds, 55
senses, and measurement, 11-15, 57
shadows, and time, 49-50, 51
shorter and longer, 13, 15, 20
small and big, 5-8
spaghetti, measuring with, 21
standard measuring units, 26, 29, 59;
for weight, 38-40;
for temperature, 43
stars, scientists and, 57-58
sterling silver, as standard unit, 38
sun, and time, 47, 48
sun dial, 49

television stations, and time, 56
temperature, 10, 41, 42, 43, 44;
record of, 61-62
thermometer, 41-46;
use of, 43;
and beans, 61-62
time, 10, 47-56
tools, for measurement, 15, 35, 41, 50, 51, 57-58, 59
traveling, and measurement, 9

units of measurement, 19, 21;
for weight, 37;
see also standard measuring units

watches, *see* clocks
water, at sea level, 44;
freezing of, 44, 45;
boiling of, 44, 45
weather, scientists and, 57
weathermen, and temperature, 46
weight, 34-40;
see also heavier and lighter
whales, 5
width, 11, 21-22
World Trade Center, 4-5

x-rays, 58

year, 47, 50, 51

zero degrees, 45, 46